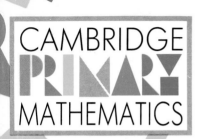

# CAMBRIDGE PRIMARY MATHEMATICS

# MODULE BOOK

## Roy Edwards
## Mary Edwards
## Alan Ward

**Cambridge University Press**
Cambridge
New York   Port Chester
Melbourne   Sydney

Published by the Press Syndicate of the University of Cambridge
The Pitt Building, Trumpington Street, Cambridge CB2 1RP
40 West 20th Street, New York, NY 10011–4211, USA
10 Stamford Road, Oakleigh, Victoria 3166, Australia

First published 1991
Reprinted 1992

Printed in Great Britain by Scotprint Ltd, Musselburgh

*British Library cataloguing in publication data*

Edwards, Roy, *1931–*
Cambridge primary mathematics.
Module 7. Bk. 1
1. Mathematics
I. Title   II. Edwards, Mary, *1936–*   III. Ward, Alan, *1932–*
510

ISBN 0 521 35325 6

The authors and publishers would like to thank the many schools
and individuals who have commented on draft material for this
course. In particular, they would like to thank Anita Straker for
her contribution to the suggestions for work with computers,
Norma Anderson, Ronalyn Hargreaves (Hyndburn Ethnic
Minority Support Service) and John Hyland (Advisory Teacher
in Tameside). They also acknowledge the Amateur Swimming
Association for permission to reproduce swimming badges.

Photographs are reproduced courtesy of:
front cover ZEFA; p 34 courtesy of the Trustees of the V & A;
p 37 Rowntree Mackintosh; pp 38–41 The Mansell Collection;
p 42 Harrowgate Museums and Art Gallery/Bridgeman Art Library
(Frith: Many Happy Returns of the Day); p 60 RSPB/J Markham;
p 61 RSPB/C H Gomersall; p 62 RSPB/G Downey;
p 63 RSPB/D Sewell; p 64 The Wildfowl & Wetlands Trust/J B Blossom

All other photographs by Graham Portlock.
The mathematical apparatus was kindly supplied by E J Arnold.

Designed by Chris McLeod

Illustrations by John Bendall Brunello
Diagrams by Oxprint
Children's illustrations by James Chatwin

# Contents

One class of children decided to make
and sell a school magazine each month.

**1** Use the price list.
Find the cost of
the children's order.
Copy and complete
the table.

| Order | cost £  p |
|---|---|
| 500 sheets of paper<br>2 staplers<br>2 boxes of staples<br>2 pots of glue | |
| total cost | |

## Price List

| | |
|---|---|
| **paper (500 sheets)** | £5-00 |
| **1 stapler** | £10-00 |
| **staples (per box)** | £3-00 |
| **glue (per pot)** | £4-50 |

What is the cost of the following numbers of sheets of paper?

**2** 100     **3** 50     **4** 10     **5** 1

One photocopy costs 5p. How much do the following numbers of photocopies cost?

| 6 | 2 | 7 | 5 | 8 | 9 | 9 | 10 |
| 10 | 20 | 11 | 25 | 12 | 60 | 13 | 100 |

**14** Copy and complete the following table.

| Cost of 1 magazine | |
| --- | --- |
| materials used | cost |
| 4 sheets of paper | ☐ p |
| 4 photocopies | ☐ p |
| staples and glue | 1p |
| Total cost | 25p |

How much do the following magazines cost?

| 15 | 2 | 16 | 4 | 17 | 8 |
| 18 | 10 | 19 | 100 | | |

## Let's investigate

Find the number of sheets of paper (not pages) in some comics or magazines that you can buy. What does each sheet of the comic cost to the nearest penny? Do they cost the same for different comics? Why do you think this is?

5

The total cost of making one magazine is 25p.

Cost includes paper, photocopies, staples and glue.

**1** Complete the table.

| Number of magazines | 1 | 10 | 50 | 100 | 150 | 200 | 250 |
|---|---|---|---|---|---|---|---|
| Cost | 25p | | | | | | |

Use your table. How much will it cost to make the following?

**2** 11 magazines   **3** 60 magazines   **4** 40 magazines

**5** 350 magazines

If magazines were sold at 30p each how much profit would be made on these sales?

**6** 10 magazines   **7** 50 magazines   **8** 100 magazines

If magazines were sold at 35p each how much profit would be made on these sales?

**9** 10 magazines   **10** 50 magazines   **11** 100 magazines

**12** If the profit on 100 magazines was £15 what was each magazine sold for?

## Cheetah Sports

A fantastic range of

## sporting clothes and equipment.

All sports catered for: Soccer, Rugby, Tennis, Swimming, Squash, Basketball, Table tennis, Lacrosse, Hockey, Darts, Athletics, Cricket, Snooker, and many more.

Mon-Fri   9:00 a.m. - 5:00 p.m.
Sat   9:00 a.m. - 1:00 p.m.

Badminton Mall, The Shopping Centre.

### ✂ Sciss ors ✂

Open 6 days a week.

9:00 a.m. - 6:00 Monday to Thursday
9:00 a.m. - 7:30 Friday and Saturday
*(An appointment is not always needed.)*

Nº 1 Blow Way

Some shopkeepers paid to advertise in the magazine.

**13**   Complete the table.

| Advertising costs | |
|---|---|
| 1 page | £10 |
| ½ page | £5 |
| ¼ page | £ ☐ |
| ⅛ page | £ ☐ |

The Bookworm Bookstore

For the very best in children's books.

40 Reading Rd.

### Pets Paradise

Everything for you and your **pet**.

The LARGEST selection of **pets** and **pet supplies** in the area.

**Pet's Paradise** - *the pet's choice.*

21 Pedigree Street

How much would these advertisements cost?

**14**   $1\frac{1}{2}$ pages    **15**   $\frac{3}{4}$ page    **16**   $1\frac{1}{8}$ pages

**17**   $1\frac{1}{4}$ pages    **18**   $\frac{5}{8}$ page    **19**   $\frac{3}{8}$ page

The Toybox

For all the toys you want at prices that will make you smile

■ have a peek in the Toybox!

*(Specialists in radio controlled toys.)*
7 Nursery Lane

## Let's investigate

Work with a group of friends. Plan a magazine that you could make. Work out the cost of the materials. Decide upon a selling price.

# C

This bar chart shows
the monthly takings
from the magazine sales.

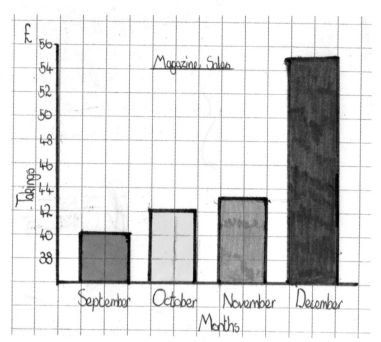

Magazine Sales

All magazines cost
25p each.

**1** Use the bar chart. Copy
and complete the table.

| Month | Takings |
|-------|---------|
| September | £40 |
| October | |
| November | |
| December | |

How many magazines were sold
in the following months?

**2** September     **3** October

**4** November     **5** December

**6** Find the total amount taken for the 4 months.

**7** Find the average amount taken for the 4 months.

**8** If £12 was also taken in advertising each month what is the total
money taken by sales and advertising together over the 4 months?

## Let's investigate

A magazine costs 40p to produce. Decide what
price you would try to sell it for. Explain the
reasons for your decision. Think of some ways
to produce the magazine more cheaply.

# *Shape 1*

Trace the handle in the photograph.
Mark one arm of your tracing.
Turn the tracing.
It fits on the photograph
in 3 different ways.

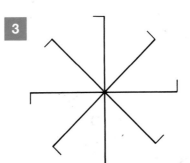

**1** It has an order of rotation of ☐ .

Use tracing paper.
Find the order of rotation of the following shapes.

**2**

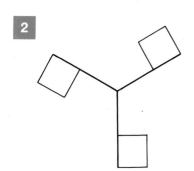

**3**

**4**

9

# Find the order of rotation of the following designs.

5

6

8

7

Find the order of rotation of the following signs.

*Let's investigate*

Work with some friends. Use a copy of the Highway Code.
Draw some road signs which have rotational symmetry.
Write their order of rotation.
Make up some more signs which have rotational symmetry.

**B**

This triangle has an
order of rotation of 3

**1** Copy the table.

| Shape | Order of rotation |
|---|---|
| equilateral triangle | 3 |
| square | |
| rectangle | |
| regular pentagon | |
| regular hexagon | |
| regular octagon | |

**2** Use tracing paper to draw round each shape. Mark one corner. Turn the tracing to find the order of rotation of the shape. Complete the table.

What is the order of rotation of the following shapes?

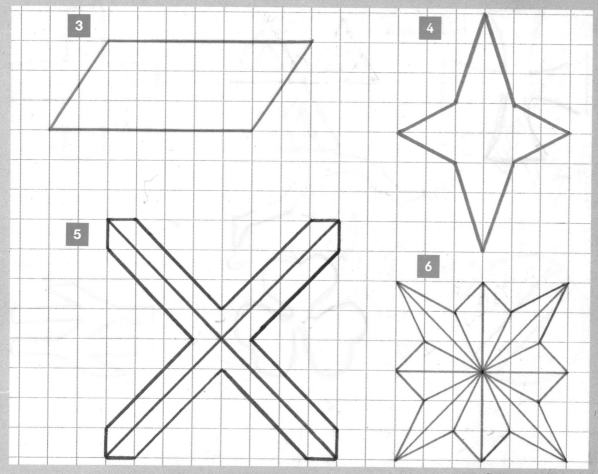

**7** Draw two different shapes of your own which have rotational symmetry.

To make a turning design:

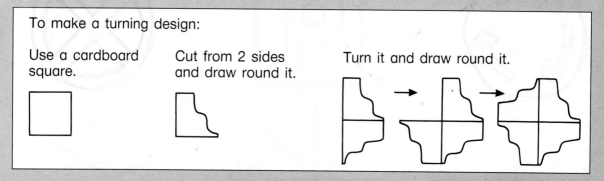

Use a cardboard square.

Cut from 2 sides and draw round it.

Turn it and draw round it.

**8** Make a turning design of your own using a cardboard square.

**9** What is the order of rotation of your design?

Find the letters in these signs which have
rotational symmetry of order more than 1.
Write their order of rotation.

**10** CHEMIST

**11** ZOO

**12** NO PARKING

**13** Copy and finish the table.
Find the order of rotation of
all the capital letters in the
alphabet.

| Order of rotation | Capital letters |
|---|---|
| 1 | A, B |
| 2 | |
| 3 or more | |

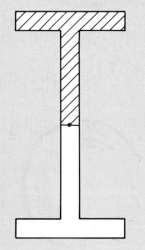

The letter T has been used to make a turning
pattern of order 2.

**14** Trace the letter T and use it to make a turning
pattern of order 4.

*Let's investigate*

Choose another capital letter. Use it to make
turning patterns of order 2 and order 4.
Try it with other letters.
Will this work for any capital letter?

**C** These patterns have been made by
turning different shapes.
Find the order of rotation of each pattern.

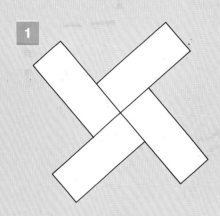

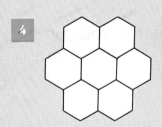

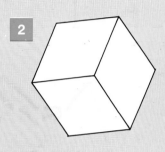

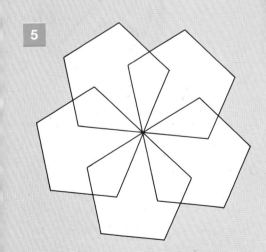

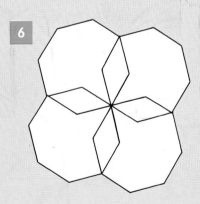

These patterns were all made by turning a square.
Find the order of rotation of each pattern.

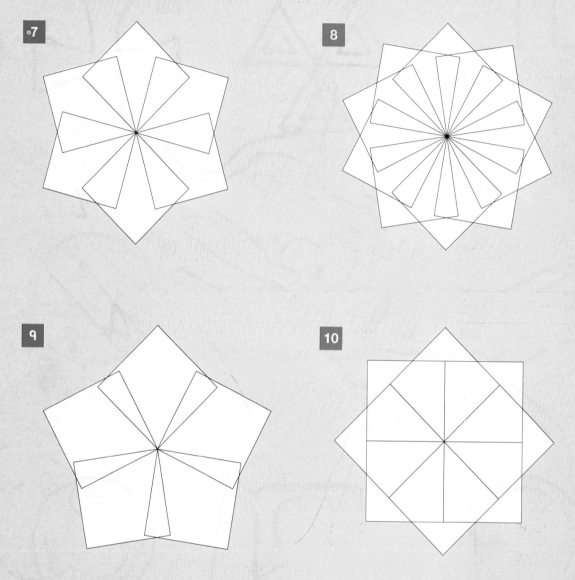

7

8

9

10

11 Make two of the square patterns yourself.
If possible use a computer.

# Let's investigate

Make other patterns which have rotational symmetry by turning a
hexagon.
Try to make different turning patterns using other shapes.

# Number 1

## A

> Each year your school orders the stock needed for your lessons. The teachers choose it from catalogues.

There are 500 cubes in a bag.

**1** Copy and complete the chart.

| Number of bags | 1 | 2 | 3 | 4 | 5 | 6 |
|---|---|---|---|---|---|---|
| Number of cubes | 500 | 1000 | | | | |

**2** How many cubes are in 10 bags?

**3** If your school needs 3000 cubes how many bags do they buy?

These cubes are shared equally among the classes.
How many does each class get?

| | Cubes | Number of classes |
|---|---|---|
| **4** | 1000 | 2 |
| **5** | 1200 | 3 |
| **6** | 1500 | 5 |
| **7** | 2000 | 4 |
| **8** | 2500 | 5 |

**9** Some bags hold 600 cubes.
How many cubes are in two of these bags?

There are 1800 art straws in each box.
How many are there altogether in these boxes?

**10**    4 boxes    **11**    5 boxes    **12**    2 boxes

**13**    How many straws in half a box?

If the 1800 straws were packed equally into other boxes
how many would there be in each box?

| | Total straws | Number of boxes | Number in each box |
|---|---|---|---|
| **14** | 1800 | 2 | ☐ |
| **15** | 1800 | 6 | ☐ |
| **16** | 1800 | 3 | ☐ |
| **17** | 1800 | ☐ | 200 |

**18**   1800 ÷ 2 = ☐
     ☐ × 2 = 1800

**19**   1800 ÷ 6 = ☐
     ☐ × 6 = 1800

**20**   1800 ÷ 3 = ☐
     ☐ × 3 = 1800

**21**   1800 ÷ ☐ = 200
     200 × ☐ = 1800

**22**    What do you notice?

**23** Use your calculator.

$12 \div 5 = \square$

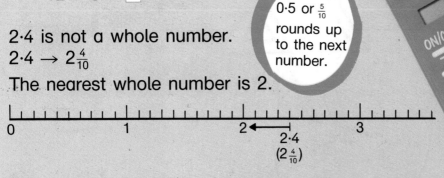

2·4 is not a whole number.

$2·4 \rightarrow 2\frac{4}{10}$

The nearest whole number is 2.

0·5 or $\frac{5}{10}$ rounds up to the next number.

Round each of these to the nearest whole number.

**24** | 1.3 |  **25** | 3.2 |  **26** | 2.9 |  **27** | 3.5 |

Do the same for these.

**28** | 8.7 |  **29** | 10.8 |  **30** | 12.4 |  **31** | 15.1 |

$9 \div 4 =$ | 2.25 | $\rightarrow 2\frac{25}{100}$

2 is the nearest whole number.

**32** $11 \div 4 =$ | 2.75 | $\rightarrow 2\frac{75}{100}$

The nearest whole number is $\square$ .

Round each of these to the nearest whole number.

**33** 3·32    **34** 4·88    **35** 6·55    **36** 8·44

## Let's investigate

Predict the answer to $8 \div 4 \times 4$. Check it on a calculator.

Predict the answer to $7 \div 3 \times 3$. Check it.

What do you notice?

Investigate dividing and multiplying other numbers by 2, 3 and 4.

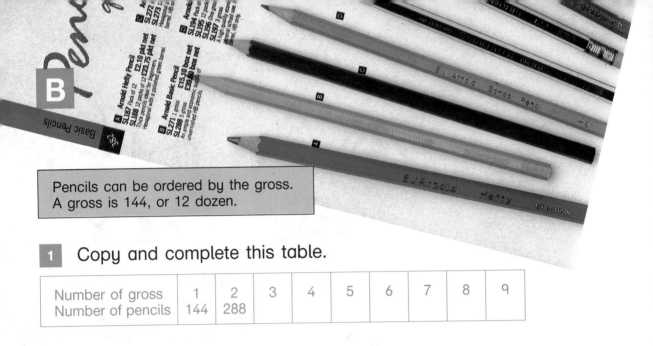

Pencils can be ordered by the gross.
A gross is 144, or 12 dozen.

**1** Copy and complete this table.

| Number of gross | 1 | 2 | 3 | 4 | 5 | 6 | 7 | 8 | 9 |
|---|---|---|---|---|---|---|---|---|---|
| Number of pencils | 144 | 288 | | | | | | | |

**2** How many pencils in 10 gross?

**3** How many pencils in 20 gross?

In the following schools the teachers ordered three pencils
a year for each child. Copy and complete the table.

| | School | Number of pupils | Number of pencils needed each year | Gross ordered each year |
|---|---|---|---|---|
| **4** | Amville | 368 | | |
| **5** | Bancroft | 279 | | |
| **6** | Canbury | 427 | | |
| **7** | Leverham | 446 | | |
| **8** | Brelton | 199 | | |

**9** Another school needs exactly 8 gross of pencils for 1 year.
They estimate each child will use 4 pencils.
How many children are there?

**10** A different school orders 10 gross for the year.
They estimate each child will need 5 pencils.
What is the largest number of children that
could be in the school?

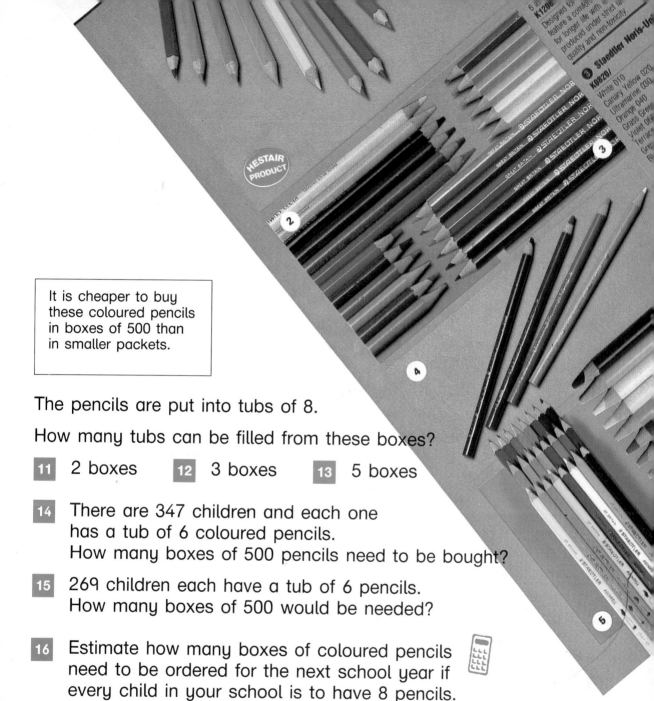

It is cheaper to buy these coloured pencils in boxes of 500 than in smaller packets.

The pencils are put into tubs of 8.

How many tubs can be filled from these boxes?

**11** 2 boxes  **12** 3 boxes  **13** 5 boxes

**14** There are 347 children and each one has a tub of 6 coloured pencils. How many boxes of 500 pencils need to be bought?

**15** 269 children each have a tub of 6 pencils. How many boxes of 500 would be needed?

**16** Estimate how many boxes of coloured pencils need to be ordered for the next school year if every child in your school is to have 8 pencils.

## Let's investigate

Find all the numbers between 0 and 200 that divide exactly by both 6 and 8. How many are there? What do you notice about them?

**C**

Paper is sometimes ordered in
reams and quires.
A ream is 480 sheets of paper.
A quire is 24 sheets.

**1** How many quires of paper
make 1 ream?

If reams of activity paper are shared
equally among the classes how many
sheets of paper does each class get?

**2** 4 reams are shared among 6 classes.

**3** 3 reams are shared among 5 classes.

**4** 5 reams are shared among 8 classes.

**5** Choose 3 classes in your school. If each child in
these classes uses approximately 6 pieces of art paper
each term, how many reams will need to be ordered
for a term? How many for a year?

**6** Estimate how many Maths exercise books
your class uses in one year.

**7** Do the same for English and Science.

**8** How many boxes will need to be ordered
for each of these subjects each year?

**9** Estimate how many Maths, English and
Science exercise books will be used by all the
Junior children in your school in one year.

*Let's investigate*

Find a way of estimating the thickness of a
single sheet of activity paper.
What about other types of paper?
Discuss your methods with a friend.

22

# Length

Architects and builders need scale plans when they are constructing buildings.
A scale plan of a building shows the true measurements drawn to a smaller scale.

**A**

**1** Measure this plan of a swimming pool.

Scale 1 cm : 1 m
or
1 cm on this plan
is 1 metre of the
swimming pool

← ———— length ———— →

**2** What is its true length?

**3** What is its true width?

How many lengths of the pool
must you swim to gain
these badges?

**4** 10 m          **5** 25 m          **6** 50 m

Use a scale of 1 cm : 1 m.
Use a ruler to draw plans for these pools.

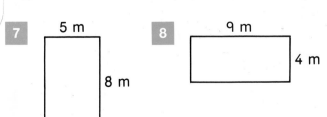

**7**
5 m

8 m

**8**
9 m

4 m

**9** Draw a plan for another pool.
Use a 1 cm : 1 m scale.
Write on the true measurements.

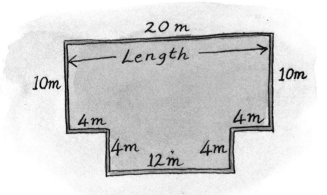

**10** Draw this pool using a
1 cm : 1 m scale.
Write the true perimeter.

How many lengths must you swim to gain these badges?

**11** 100 m badge    **12** 200 m badge

**13** Jim swims 6 lengths. How many more metres
must he swim to gain his 200 m badge?

**14** Use a 1 cm : 2 m scale.
Draw a plan for this swimming pool.

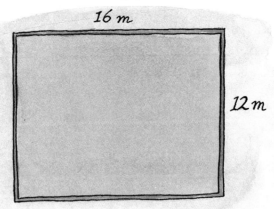

# Let's investigate

Use a 1 cm : 1 m scale.
Draw plans for pools with a perimeter of 40 m.
Each pool must be different.

This rectangular swimming pool is 25 m long and 12 m wide. It has 6 swimming lanes.

**1** Use a 1 cm : 2 m scale.
Draw a plan of the pool and show the lanes.

**2** To gain this award you have to swim 800 m.
How many lengths of the 25 m pool is this?

**3** You must swim using 3 different
strokes for at least 200 m in each.
Suggest some different ways of
swimming the 800 m.

Look at the diagrams of the two pools below.
Draw scale plans of them. Choose your own scale.
Write on the missing measurements.

**4**

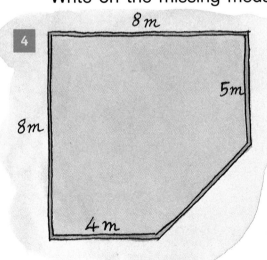

**5**

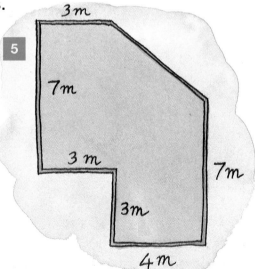

This is a scale drawing of part of a sports centre.
The scale is 1 cm : 2 m.

**6** Measure then draw an accurate plan.
Write on the true measurements.

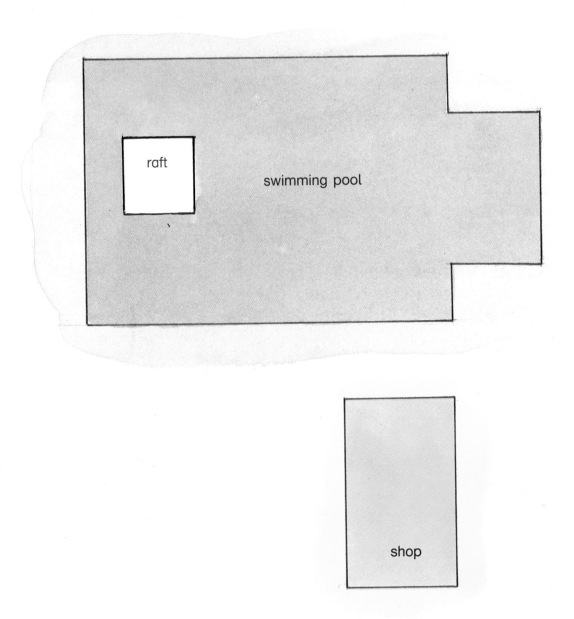

**7** Find somewhere on your scale plan to add
a paddling pool 3 m by 2 m.

# Let's investigate

Measure your school hall to the nearest metre.

Design a swimming pool that would fit in the hall.
Don't forget to leave an area to walk round.

Draw a scale plan of your hall with the swimming pool in it.
Choose the best scale for it.

## C  Let's investigate

A garden pool and
a lake in a park are
both this shape,
but very different sizes.
Decide on some true measurements for each of the two pools.

Choose a suitable scale for drawing a plan of each one.
Draw the two scale plans as accurately as you can.
Explain how you did it.

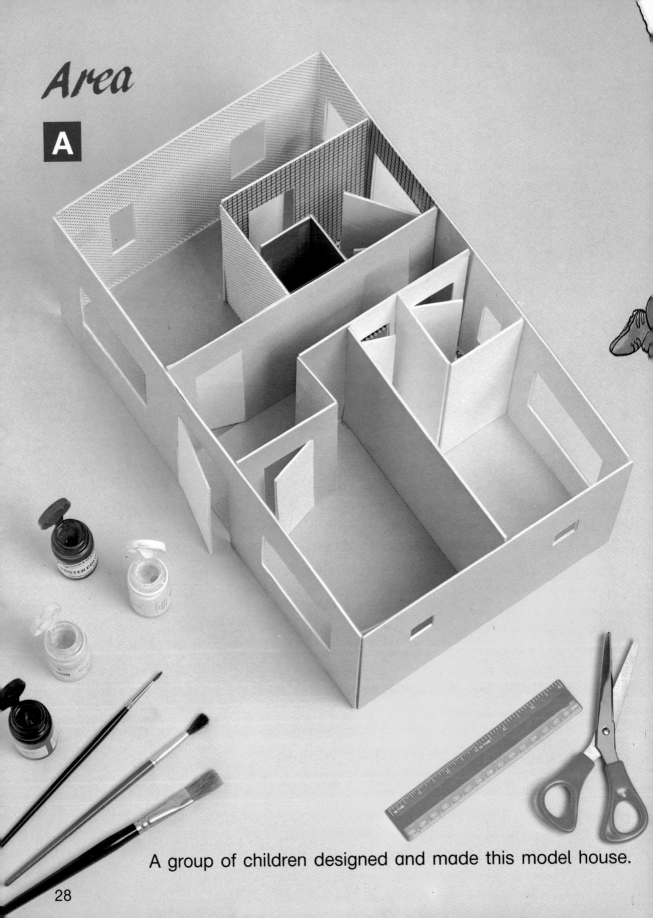

A group of children designed and made this model house.

1. The area of the dining room is ☐ cm².

2. The area of the living room is ☐ cm².

Felt is being used to carpet the rooms.
How much felt is needed for these rooms?

3. The dining room.

4. The living room.

5. Both the dining room and the living room.

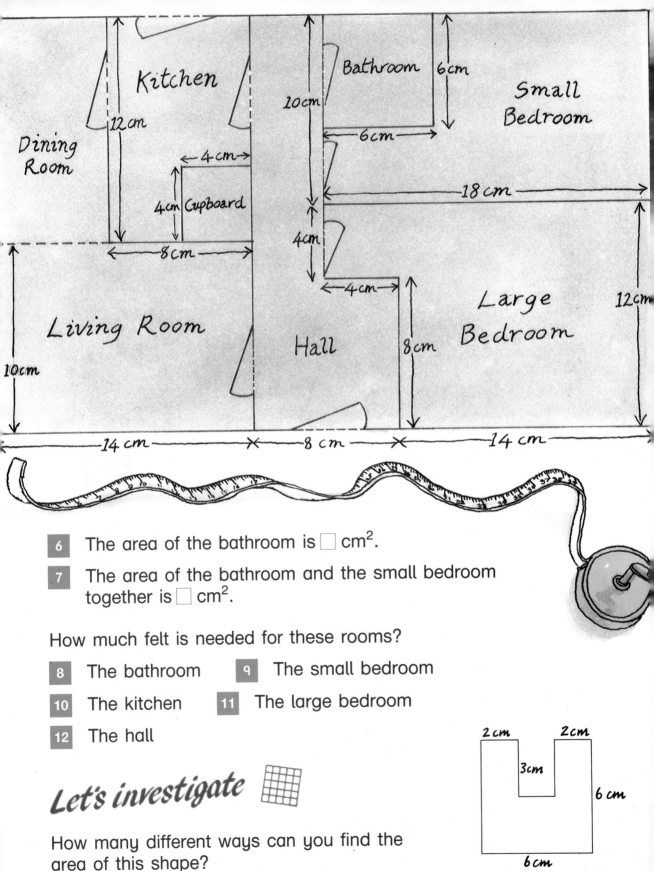

6    The area of the bathroom is ☐ cm².

7    The area of the bathroom and the small bedroom together is ☐ cm².

How much felt is needed for these rooms?

8    The bathroom     9    The small bedroom

10    The kitchen     11    The large bedroom

12    The hall

# Let's investigate

How many different ways can you find the area of this shape?

30

## B

The doors of the model house need
another coat of paint.
Each area needs to be painted
in its original colour.

1  The Red area on this door is ☐ cm².
2  The Yellow area on this door is ☐ cm².

Complete these.

3
Blue
area = ☐ cm²

Green
area = ☐ cm²

4
Red
area = ☐ cm²

White
area = ☐ cm²

5
White
area = ☐ cm²

Black
area = ☐ cm²

6
Orange
area = ☐ cm²

Green
area = ☐ cm²

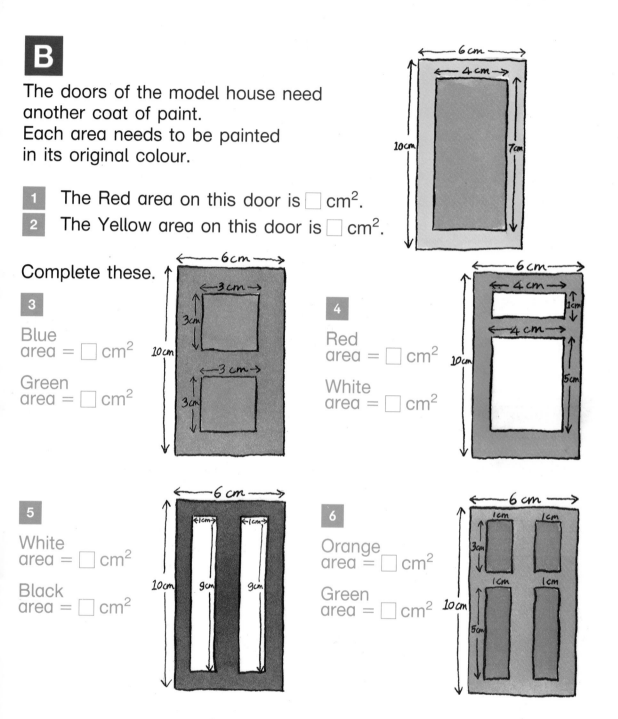

# Let's investigate

What is the smallest area you can make with a perimeter of 16 cm?
What is the largest area you can make with a perimeter of 16 cm?

**C**

13 cm

36 cm

22 cm

13 cm

The walls of the
model house
need a coat of
paint.

13 cm

22 cm

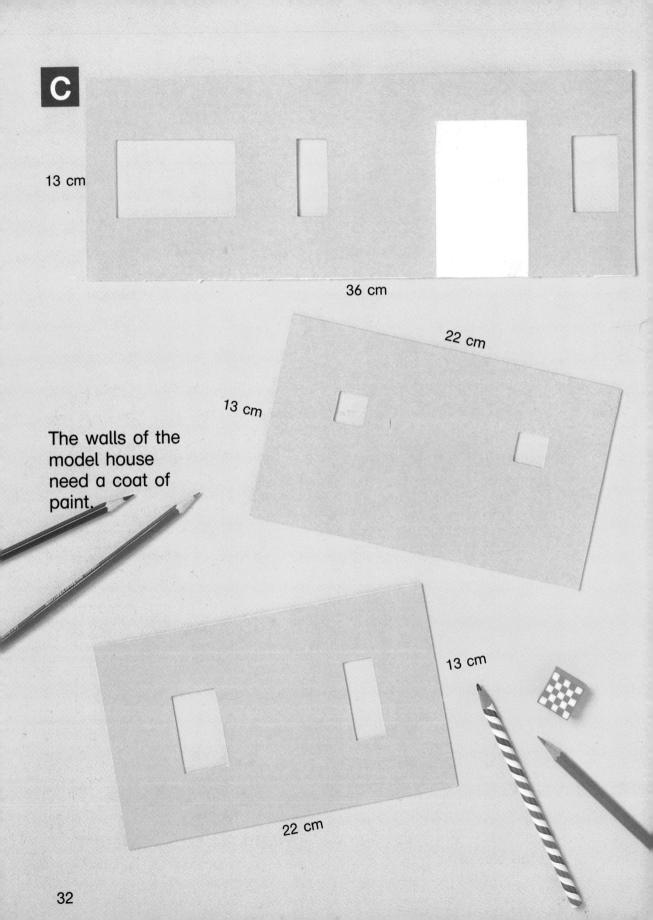

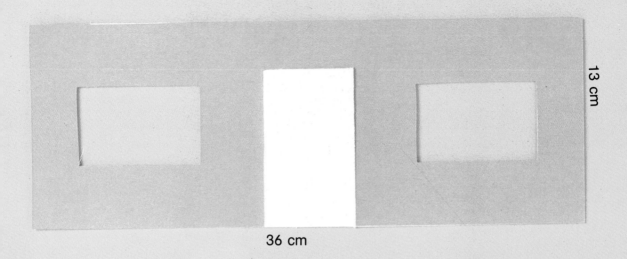

13 cm

36 cm

Look at the measurements on all the walls.

**1** Approximately what area needs to be painted?

**2** How many pots of paint will you need?

One pot of paint will just cover 1300 cm²

This chart gives measurements of the windows and doors.

**3** Copy and complete it.

**4** Use the chart to find the exact area of wall that needs to be painted.

**5** How many pots of paint will be needed?

| Height | Width | Area |
|--------|-------|------|
| ☐ cm | 8 cm | 40 cm² |
| ☐ cm | 2 cm | 4 cm² |
| 2 cm | 2 cm | ☐ cm² |
| 5 cm | ☐ cm | 40 cm² |
| 5 cm | ☐ cm | 15 cm² |
| ☐ cm | 3 cm | 15 cm² |
| 5 cm | 8 cm | ☐ cm² |
| ☐ cm | 2 cm | 10 cm² |
| 5 cm | 2 cm | ☐ cm² |
| 10 cm | ☐ cm | 60 cm² |
| 10 cm | 6 cm | ☐ cm² |

*Let's investigate*

Draw accurately a rectangle 5 cm long and 4 cm wide.
Divide it into four different shapes, each with the same area.
Find other ways of doing it.

# Volume

## A

The volume of a tin or box is
the amount of space it contains.
It may be measured in
cubic centimetres.

**1** Make different cuboids.
Use 12 centimetre cubes
to make each one.
Record their measurements.

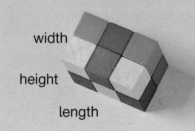

width

height

length

| Length | Width | Height | Volume |
|--------|-------|--------|--------|
| cm | cm | cm | 12 cm³ |

**2** What do you notice about the measurements and the volume?

Use cubes to make these shapes. Find their volumes.
Record their measurements in your table.

**3**
width 2 cm
height 2 cm
length 2 cm

**4**
2 cm
2 cm
3 cm

**5**
3 cm
2 cm
4 cm

# Find the volumes of these shapes.

Many people collect old printed tin boxes. In 1971 an exhibition of biscuit tins was held at the Victoria and Albert Museum in London.

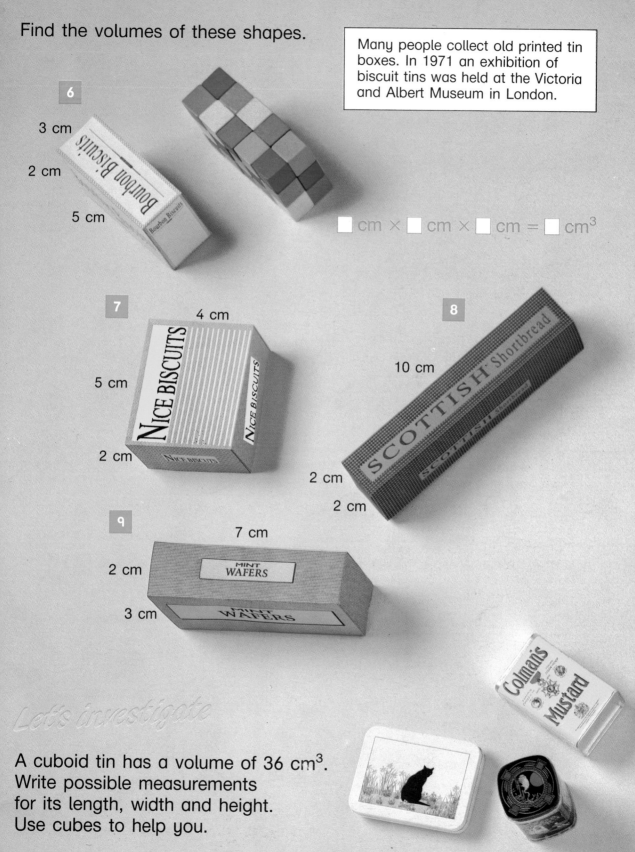

**6**

3 cm

2 cm

5 cm

☐ cm × ☐ cm × ☐ cm = ☐ cm³

**7**

4 cm

5 cm

2 cm

**8**

10 cm

2 cm

2 cm

**9**

7 cm

2 cm

3 cm

*Let's investigate*

A cuboid tin has a volume of 36 cm³.
Write possible measurements
for its length, width and height.
Use cubes to help you.

**B**

**1** The volume of the cube is 8 cm³.

2 cm × 2 cm × ☐ cm = 8 cm³

What is the height?

Find the missing measurements.

**2**

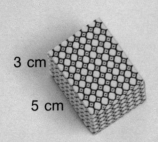

3 cm

5 cm

☐ cm × 3 cm × 5 cm = 60 cm³

**3**

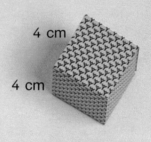

4 cm

4 cm

4 cm × ☐ cm × 4 cm = 64 cm³

**4**

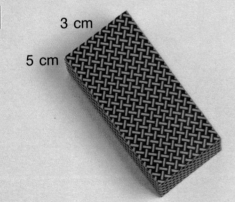

3 cm

5 cm

☐ cm × 3 cm × 5 cm = 150 cm³

**5**

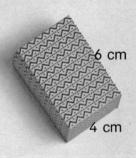

6 cm

4 cm

6 cm × 4 cm × ☐ cm = 72 cm³

**6** Copy and complete the table.

| Length | Width | Height | Volume |
|--------|-------|--------|--------|
| 3 cm | ☐ cm | 1 cm | 6 cm³ |
| ☐ cm | 5 cm | 4 cm | 120 cm³ |
| 6 cm | 3 cm | ☐ cm | 36 cm³ |
| 8 cm | ☐ cm | 2 cm | 80 cm³ |

36

## Let's investigate

Use cubes to make the next 3
larger cubes.
Make a chart to show the
measurements and the volumes.
Predict how many cubes you need for the next two larger cubes.
What pattern do you find in the measurements?

1 cm³    8 cm³

## C

Some old tins were very unusual
shapes. One small tin was shaped
like a Christmas pudding on a plate.
The pudding shape lifted off the
plate to open the tin.

1   If the diameter of the plate is 10 cm
    and the height of the pudding is 4 cm,
    design a cuboid box that the tin
    just fits into.
    Write the measurements and volume of the box.

2   Make the box using squared paper.

3   Choose any object in the classroom.
    Design and make a cuboid shaped box to pack it in.
    Write the measurements and volume of the box.

## Let's investigate

Collect some soap powder boxes.
You need E3 and E15 sizes.
Find their volumes.
What do you discover?
Investigate other soap box sizes.

SAINSBURY'S
WASH & CARE
AUTOMATIC
BIOLOGICAL
WASHING POWDER
E3 size 1.2 Kg

# Probability 1

**A**

| 0 | | | | 1 |
|---|---|---|---|---|
| No chance | Poor chance | Even chance | Good chance | Certain |

Draw the scale.
Draw arrows to show the chance or likelihood of these things happening.
Look at the picture for clues. It shows a Victorian classroom.

1. The benches were uncomfortable to sit on.

2. There was a computer in the classroom.

3. The children wrote in ink.

4. Victorian children learnt the two times table.

5. There were lots of exciting books to look at.

6. The teacher's age was an odd number.

7. The school was in one room.

8. The school had central heating.

9. Explain your answers to a friend.
   Does your friend get the same answers?

```
0                                  1
├──────────────┴──────────────┤
No chance  Poor chance  Even chance  Good chance  Certain
```

Copy the scale. Draw arrows for this picture.

**10** The boys enjoyed their P.E. lesson.

**11** Were you sure where to put the arrow? Explain why.

Show where you think the following fit on the probability scale.

**12** Boys and girls did P.E. together.

**13** The boys changed into shorts and trainers.

**14** They all did the same exercise at the same time.

**15** The next exercise was to put their arms at their sides.

**16** They marched back into school in lines.

## Let's investigate

Work with a friend. Think of a P.E. lesson in school.
Write about what could happen and show
the chances on a probability scale.

**B**

Work with a friend.
Put these sentences about poor Victorian
children on a probability scale.

1   Poor children started work at a young age.

2   Children were employed because they were
small enough to go down narrow tunnels.

3   They grew up strong and healthy.

4   They got injured at work.

5   Poor children came from rich homes.

6   They were treated kindly at work.

7   They worked long hours.

8   They earned a lot of money.

9   Make up three more sentences about children
working in Victorian times.
Write them and show them on the scale.

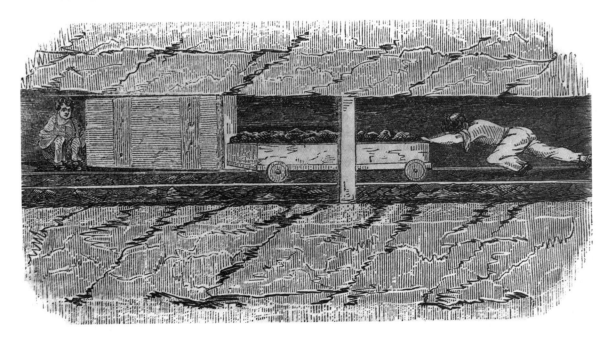

```
     0    1/10   2/10   3/10   4/10   5/10   6/10   7/10   8/10   9/10    1
```

No
chance

Even
chance

Certain

Write a whole number or fraction to show what you
think was the chance of these things happening in
Victorian coach travel. Give reasons for your answer.

**10** There was a petrol station at the end of the road.

**11** The people on top would get wet if it rained.

**12** The journey might be bumpy.

**13** This was a London coach.

**14** The driver had a good view of the road.

**15** It would be noisy on the coach.

# Let's investigate

The likelihood of there being traffic jams at holiday times today is $\frac{9}{10}$.
Write whether you agree or disagree. Give reasons.

Make up some probability fractions for bus, train
or car travel today.
Explain why you chose the fractions.

C

$$0 \qquad \frac{1}{2} \qquad 1$$

Copy this scale. Draw arrows to show these probabilities.

1. The next child to be born in the family will be a girl.

2. Victorian families had equal numbers of girls and boys.

3. Show the probability that the mother's birthday is in September.

4. Show the probability that the father's last birthday was on a Monday.

## Let's investigate

Plan other scales and show probability fractions on them. Explain what you have done.

# Map reading

Look at the grid.
The numbers going from west to east are **eastings**.
The numbers going from south to north are **northings**.

**A**

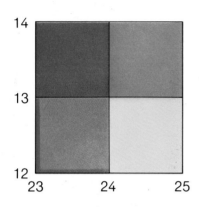

1  Write the eastings.

2  Write the northings.

Look at the green square.
Its easting is 23. Its northing is 12.
Its grid reference is 23 12.

3  What colour square is 24 13?

What are the grid references for these squares?

4  blue      5  yellow

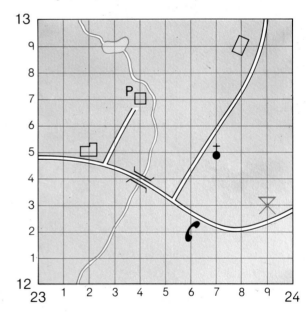

The exact easting for
the church is 237.
The exact northing for
the church is 125.

The 6 figure grid reference
for the church is 237 125.

Write 6 figure grid references for these places.

6  bridge ⋈

7  post office P

8  building ▭

9  telephone box ☎

10  school ⌐▯

11  picnic site ⧓

12  pond ⬭

Draw the four large squares of this grid on squared paper.

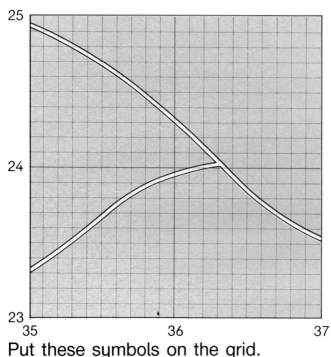

**Key**

| | |
|---|---|
| P | Post Office |
| + | Church |
| ♦ | Church with a spire |
| ■ | Church with a tower |
| ▭ | Bus station |
| P | Parking |
| i | Tourist Information |
| ⚑ | Golf course |
| ✕ | Picnic site |
| Λ | Camp site |
| ⚏ | Caravan site |
| ☎ | Telephone |
| ⚔ | Battle site |

Put these symbols on the grid.

**13** Post Office at 354 237

**14** Camp site at 358 232

**15** Tourist Information at 365 241

**16** Church with a spire at 353 247

**17** Bus station at 368 235

**18** Golf course at 365 245        **19** Caravan site at 359 249

**20** Choose two more symbols from the list.
Put them on the grid. Write their grid references.

1 centimetre on Map A is 25 metres on the ground.

Make this ruler.
Use it to measure these lines.
How many metres on the ground would they be?

| 0 | 25 | 50 | 75 | 100 | 125 | 150 | 175 | 200 |
|---|---|---|---|---|---|---|---|---|

metres

**21** _____

**22** _____

What is the distance in metres as the crow flies between these places?

23   the church with the tower and the church with the spire

24   the bridge over the stream and Hills Farm

25   the pond and the church with a spire

26   the Post Office and the bridge

Find these places and write their names.
Write the 6 figure reference for each one.

27   75 m S of 𝄐

28   125 m E of Hills Farm

29   25 m SE of ⛫

30   100 m SW of the Post Office

## Let's investigate

Find a way to measure the distance by road and path from Hills Farm to the pond.
Explain how you did it.

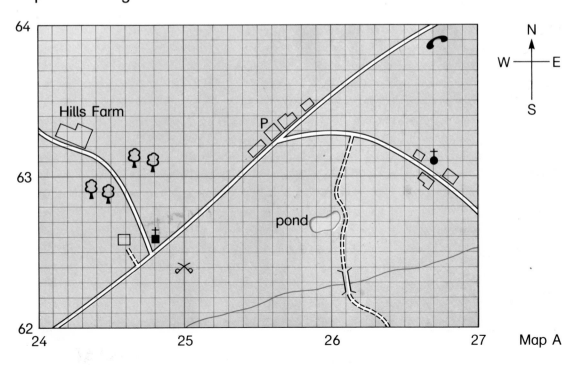

Map A

## B

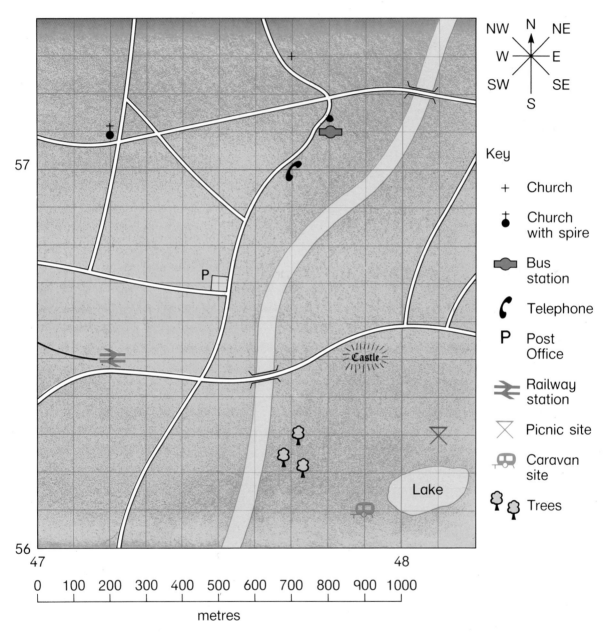

Key

+    Church

♀    Church with spire

🚌    Bus station

📞    Telephone

P    Post Office

🚉    Railway station

✕    Picnic site

🚐    Caravan site

🌳 🌳    Trees

0   100   200   300   400   500   600   700   800   900   1000

metres

1 centimetre on the map is 100 metres on the ground.

What do you find at these references?

| | | | | | | | |
|---|---|---|---|---|---|---|---|
| **1** 477 570 | | **2** 477 573 | | **3** 481 563 | | **4** 479 565 | |
| **5** 472 565 | | **6** 478 571 | | **7** 479 561 | | **8** 475 567 | |

46

Find the difference between the distance as the crow flies
and the distance by road.

9  from  +  to  ☎

10  from  ⚲  to  🚗

11  from  ⇌  to  P

12  from castle to  P

Distance as the
crow flies =  ☐ m
Distance by road = ☐ m
Difference =  ☐ m

Follow these directions for a walk.

13  Start at 481 563. Where are you?

14  Go SW for 300 m. Where are you now?

15  Look NW. What do you see?

16  Go N for 400 m. Where are you?

17  Approximately how far is it by road
from here to the railway station?

Use the same scale as on the map.
Draw lines to show these distances.

18  850 m

19  1000 m

20  1200 m

21  250 m

*Let's investigate*

The shortest distance between a telephone
box and a post box is 200 m.
Choose a scale and draw a line to show
this distance.
Use different scales to show this distance.
What do you notice about the scales and
the lengths of the lines?

**Key**

| | | | | | |
|---|---|---|---|---|---|
| ═══ | road | 〇 | quarry | + | church |
| ══ | narrow road | 〰〰〰 | crags | ⋀ | camp site |
| ===== | track | ✲ | marsh | 🚐 | caravan site |
| - - - - | footpath | 🌲 🌲 | coniferous trees | P | car park |
| | | | | ▲ | youth hostel |

1 centimetre on the map represents 250 metres on the ground.

Follow the walk and answer the questions.
Start at the crossroads at 272 573.

**1** Follow the road in an easterly direction.
Cross White Beck.
Write the grid reference of the crossing.

**2** How many metres is it from here to the first
footpath on your left?

**3** Go north along this footpath.
What sort of ground is on the west of the path after about 250 m?

**4** Continue along the path until you reach the youth hostel.
Give the grid references of five things that you pass
on your walk.

Find the distances in metres between the following places by
road or path. Choose the shortest distance where possible.

**5** the church and the boat house

**6** the church and the waterfall

**7** the hotel and the caravan site

**8** the village crossroads and the youth hostel

**9** How far is it by canoe from the canoe centre to Green Island?

**10** Find some features that tell you this is a tourist area.
Say what they are and write their grid references.

## Let's investigate

Look at an Ordnance Survey map with a scale of 1 cm : 250 m.
Plan a walk of approximately 6 km.
Describe the walk and the features to be seen.
Give grid references and directions for your walk.

*Data 1*

**A**

1. How many different kinds of animal are in the picture?

2. Explain how you can sort the animals into two groups. List each group.

3. Sort each group with more than one animal into two more groups.

4. Keep sorting each new group until each animal is in a group by itself.

Here is a way to sort the animals into groups by themselves.
It is called a decision tree diagram.

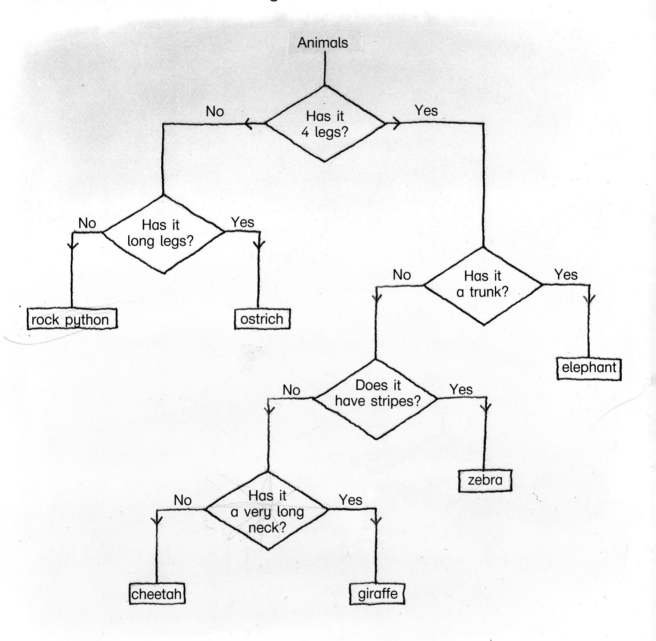

5   Play with a friend.
    A player chooses one of the animals. The
    other has to find which one it is by
    asking the questions in the decision tree diagram.

6   Take turns until you have identified all the animals.

**7** Play with a friend.
Take turns to choose one of these pictures.
The other player has to find which one it is using the
questions in the decision tree diagram opposite.

Jemba Cliff

ASAD KHAN

Jim Perkins

TONY SHAW

ULLA BERG

Tina choi

JOSIE HILL

SABA KABUR

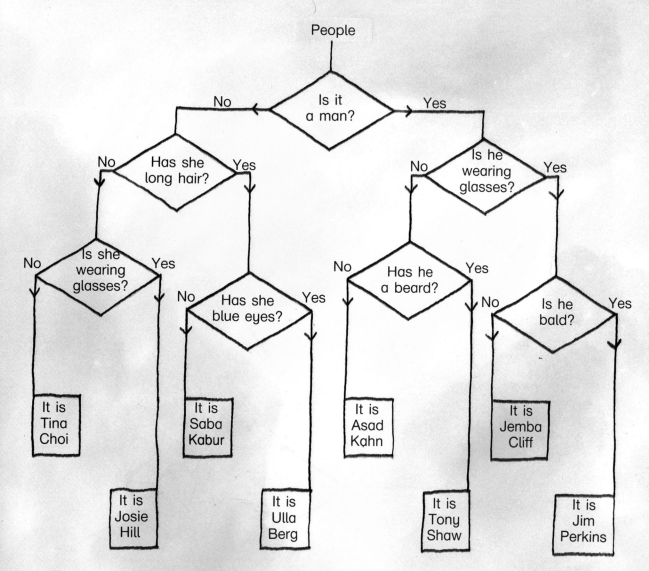

**8** Choose a different person and use the decision tree again to find him or her.

Use the decision tree to find the names of these people.
Write the questions that you answer.

**9** a bald man with glasses

**10** a long haired woman with blue eyes.

## Let's investigate

Draw the faces of four different people.
Make up your own decision tree to identify them.
Ask a friend to check your decision tree.

The answers to the questions must be yes or no.

daddy-long-legs

butterfly

dragonfly

fly

grasshopper

earwig

flea

ant

54

Look at the picture. Use it to put the missing names
of the different insects into the decision tree diagram.

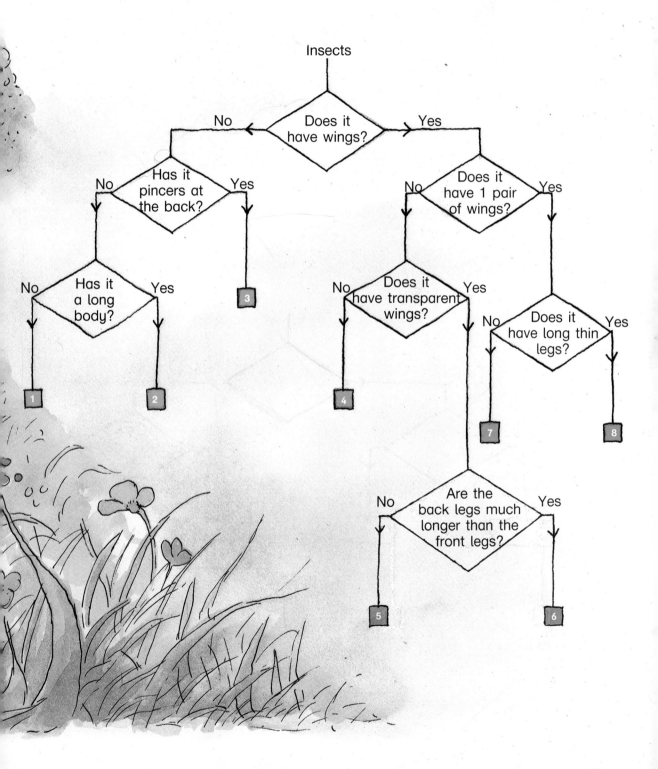

Insects

Does it
have wings?

No — Has it
pincers at
the back?

Yes

Does it
have 1 pair
of wings?

No — Has it
a long
body?

Yes

No

Yes

Does it
have transparent
wings?

No

Yes

Does it
have long thin
legs?

No

Yes

1

2

3

4

7

8

Are the
back legs much
longer than the
front legs?

No

Yes

5

6

A decision tree diagram has been used to sort the small garden animals.

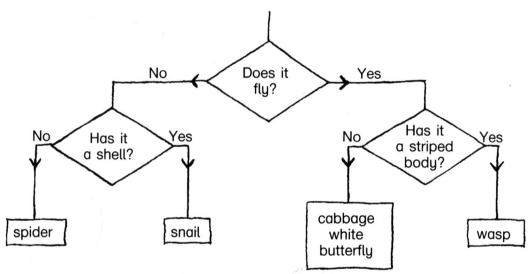

**9** Use the decision tree diagram to complete the table.

| garden animal | description |
|---|---|
| snail<br>cabbage white butterfly<br>wasp<br>spider | It has a ____ and it does not ____. |

These animals have been sorted into different groups.

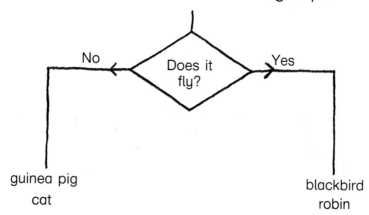

No     Does it fly?     Yes

guinea pig
cat

blackbird
robin

**10** Finish the decision tree diagram so that each animal is in a group of its own.

**11** Describe each animal using your decision tree.

## Let's investigate

Choose some different animals.
Complete a decision tree diagram
for them.
Ask a friend to check it.

The information about each person can be put onto a database like this.

Name: Zillah
Hair colour: red
Face shape: round
Height: tall

Name: Joan
Hair colour: red
Face shape: thin
Height: medium

Name: Rita
Hair colour: black
Face shape: thin
Height: medium

Name: Sarah
Hair colour: red
Face shape: thin
Height: short

Name: Mandy
Hair colour: brown
Face shape: round
Height: medium

Database

**1** Use the database. Name the 5 women in the identity parade.

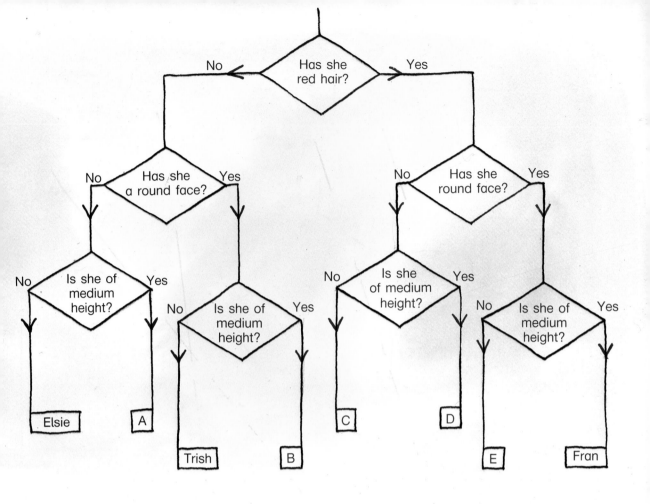

**2** Use the decision tree and database to name the following people.

A is _____  B is _____  C is _____

D is _____  E is _____

**3** Copy and complete a database for these women in the decision tree.

Name: *Elsie*
Hair colour:

Name: *Trish*
Hair colour:

Name: *Fran*
Hair colour:

## Let's investigate

Make a database for some children in your class.
Draw a decision tree for the data and use
a letter instead of each name. Ask a friend
to find out which letter stands for each name.

# Percentages

## A

Some children made a tally chart of 100 birds
that visited their playground.

| Name of bird | blackbird | gull | pigeon | sparrow | starling | thrush |
|---|---|---|---|---|---|---|
| Number | III | ＪＨＴ ＪＨＴ ＪＨＴ ＪＨＴ ＪＨＴ ＪＨＴ ＪＨＴ ＪＨＴ II | ＪＨＴ ＪＨＴ ＪＨＴ II | ＪＨＴ ＪＨＴ ＪＨＴ ＪＨＴ ＪＨＴ | ＪＨＴ ＪＨＴ II | I |

**1** Make a table to show the totals.

| Name of bird | blackbird | gull | pigeon | sparrow | starling | thrush |
|---|---|---|---|---|---|---|
| Number | 3 | | | | | |

**2** Check that the numbers add up to 100.

The fraction of blackbirds is $\frac{3}{100}$.

**3** $\frac{\square}{100}$ were starlings.

**4** $\frac{\square}{100}$ were gulls.

The children visited a park.
This is a tally chart of 100 birds they saw there.

| Name of bird | blackbird | gull | pigeon | robin | sparrow | starling |
|---|---|---|---|---|---|---|
| Number | II | ＪＨＴ ＪＨＴ ＪＨＴ ＪＨＴ ＪＨＴ IIII | ＪＨＴ ＪＨＴ | II | ＪＨＴ ＪＨＴ ＪＨＴ ＪＨＴ ＪＨＴ III | ＪＨＴ ＪＨＴ ＪＨＴ ＪＨＴ ＪＨＴ ＪＨＴ IIII |

**5** Make a table to show these totals.

The Royal Society for the Protection of Birds is known as the R.S.P.B. It protects and cares for wild birds. Its symbol is the avocet.

% or percent means 'out of 100'.

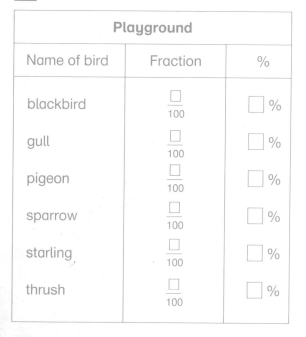

avocet

$\frac{2}{100}$ of the birds were blackbirds.
We say that 2% were blackbirds.

**6** Copy and complete the tables.

| Playground | | |
|---|---|---|
| Name of bird | Fraction | % |
| blackbird | $\frac{\square}{100}$ | $\square$ % |
| gull | $\frac{\square}{100}$ | $\square$ % |
| pigeon | $\frac{\square}{100}$ | $\square$ % |
| sparrow | $\frac{\square}{100}$ | $\square$ % |
| starling | $\frac{\square}{100}$ | $\square$ % |
| thrush | $\frac{\square}{100}$ | $\square$ % |

| Park | | |
|---|---|---|
| Name of bird | Fraction | % |
| blackbird | $\frac{\square}{100}$ | $\square$ % |
| gull | $\frac{\square}{100}$ | $\square$ % |
| pigeon | $\frac{\square}{100}$ | $\square$ % |
| robin | $\frac{\square}{100}$ | $\square$ % |
| sparrow | $\frac{\square}{100}$ | $\square$ % |
| starling | $\frac{\square}{100}$ | $\square$ % |

**7** Where was the highest fraction of birds seen?

**8** What was the highest percentage of birds?

**9** What were the birds?

**10** Where was the smallest fraction of birds seen?

**11** What was the smallest percentage of birds?

**12** What were the birds?

In another tally of 100 birds

$\frac{1}{2}$ were starlings, $\frac{1}{4}$ were sparrows and $\frac{1}{4}$ were gulls.

How many birds of each kind were there?

**13** ☐ starlings   **14** ☐ sparrows   **15** ☐ gulls

**16** $\frac{\square}{100}$ or ☐ % were starlings.   **17** $\frac{\square}{100}$ or ☐ % were gulls.

**18** $\frac{\square}{100}$ or ☐ % were sparrows.

## Let's investigate

Find fractions that are the same as 50%.

## B

200 birds were tallied on an estuary.

$\frac{1}{2}$ were terns, $\frac{1}{4}$ were gulls, $\frac{1}{4}$ were redshanks.

**1** Copy and complete the table.

| Bird | Total seen | Number out of 200 | Number out of 100 | % |
|------|-----------|-------------------|-------------------|---|
| tern | ☐ → | $\frac{\square}{200}$ → | $\frac{\square}{100}$ → | ☐ % |
| gull | ☐ | $\frac{\square}{200}$ | $\frac{\square}{100}$ | ☐ % |
| redshank | ☐ | $\frac{\square}{200}$ | $\frac{\square}{100}$ | ☐ % |

**2** What was the largest number of birds seen?

**3** What fraction was it? $\frac{\square}{100}$   **4** What percentage was it?

200 people visited a bird reserve.
How many people did the following?

**5** 60% visited the café for a drink.

**6** 55% bought something from the gift shop.

**7** 35% bought a guide book.

**8** 90% arrived by car.

**9** 45% visited one of the hides.

**10** 75% ate a picnic at the reserve.

**11** What percentage of visitors did not buy a guide book?

**12** Why do you think such a big percentage of visitors arrived by car?

**13** 40% of the visitors were children.
How many adults were there?

## Let's investigate

Work with a friend.
Find 50%, 25% and 10% of different numbers less than 100.
Record your findings.

Young bird watchers can join the Young Ornithologists' Club or the Y.O.C. It started in 1965. Its symbol is a kestrel.

Sir Peter Scott founded the Wildfowl Trust Centres. The first one was at Slimbridge on the banks of the River Severn in Gloucestershire.

**1** A school party of 50 visited a wildfowl centre.
50% were girls.
10% were adults.
How many were boys?

**2** 50% of the birds Winston saw were mallards.
25% of the birds Tina saw were mallards.
10% of the birds Raj saw were mallards.
They each saw 10 mallards.
How many birds altogether did each see?

## Let's investigate

In a flock of birds 40 were nesting.
Find some possible percentages for the nesting birds.

☐ % of ☐ = 40

How many different ways can you find?
Discuss them with a friend.